Caribou Migration

by Grace Hansen

ANIMAL MIGRATION

Abdo Kids Jumbo is an Imprint of Abdo Kids
abdobooks.com

abdobooks.com

Published by Abdo Kids, a division of ABDO, P.O. Box 398166, Minneapolis, Minnesota 55439. Copyright © 2021 by Abdo Consulting Group, Inc. International copyrights reserved in all countries. No part of this book may be reproduced in any form without written permission from the publisher. Abdo Kids Jumbo™ is a trademark and logo of Abdo Kids.

Printed in the United States of America, North Mankato, Minnesota.

052020

092020

Photo Credits: Alamy, iStock, National Geographic Image Collection, Shutterstock

Production Contributors: Teddy Borth, Jennie Forsberg, Grace Hansen
Design Contributors: Dorothy Toth, Pakou Moua

Library of Congress Control Number: 2019956467
Publisher's Cataloging-in-Publication Data

Names: Hansen, Grace, author.
Title: Caribou migration / by Grace Hansen
Description: Minneapolis, Minnesota : Abdo Kids, 2021 | Series: Animal migration | Includes online resources and index.
Identifiers: ISBN 9781098202309 (lib. bdg.) | ISBN 9781098203283 (ebook) | ISBN 9781098203771 (Read-to-Me ebook)
Subjects: LCSH: Caribou--Juvenile literature. | Caribou--Behavior--Juvenile literature. | Reindeer—Behavior--Juvenile literature. | Animal migration--Juvenile literature. | Animal migration--Climatic factors--Juvenile literature.
Classification: DDC 599.658--dc23

Table of Contents

Caribou 4

Great Group Migration 8

Back to the Summer Home 18

Western Arctic Caribou Herd
Migration Route 22

Glossary 23

Index 24

Abdo Kids Code. 24

Caribou

Caribou are mainly found in Alaska and Canada. They live in the **tundra** and **taiga** areas.

Caribou are built to survive in cold and harsh conditions. However, their greatest means of survival is **migration**.

Great Group Migration

Caribou **migrate** between summer and winter ranges. They are almost always on the move. Caribou herds can travel up to 3,000 miles (4,800 km) a year.

9

Their summer range has **nutritious** food. Eating these plants allows mothers to make rich milk. This helps the calves grow faster and stronger.

11

The summer range also has fewer **predators** and insects. It is an ideal place for raising young.

However, the summer range is harsh in the winter months. Strong winds and very cold temperatures make it unlivable. So the caribou move south between August and October.

Caribou's winter range has more food and is not as cold. The herd will remain there until late March.

Back to the Summer Home

In early April, they will begin their journey north again. Pregnant females start moving first.

In early June, mothers give birth. The rest of the herd makes it back to the summer grounds by mid-July. Soon, their **migration** south will begin again.

Western Arctic Caribou Herd Migration Route

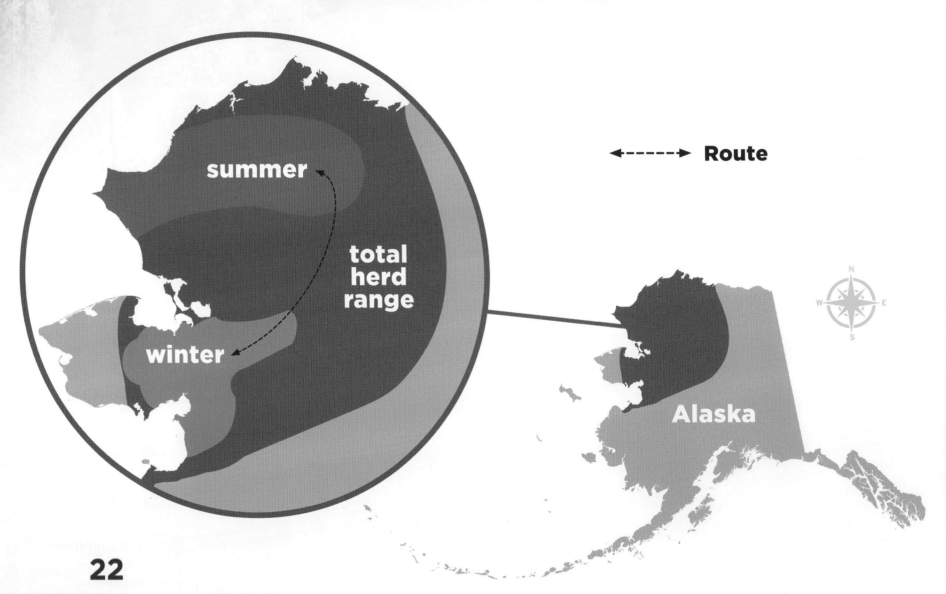

Glossary

migrate – to move from one place to another for food, weather, or other important reasons.

migration – the act or process of migrating.

nutritious – having a lot of vitamins, minerals, or other nutrients necessary to grow or stay strong.

predator – an animal that hunts other animals for food.

taiga – the strip of subarctic evergreen forest that covers much of the northern parts of North America, Europe, and Asia.

tundra – one of the huge plains in the arctic regions of North America, Europe, and Asia. Trees do not grow on tundra.

Index

Alaska 4

babies 10, 12, 20

Canada 4

fall 14

females 10, 18, 20

food 10, 16

habitat 4

herd 8, 16, 20

spring 16, 18

summer 8, 10, 12, 14, 20

winter 8, 14, 16

Visit **abdokids.com** to access crafts, games, videos, and more!

Use Abdo Kids code **ACK2309** or scan this QR code!